The **1333** Most Frequently Used **ELECTRICAL** Terms

Diccionario de Términos Eléctricos

English-Spanish-English

Inglés-Español-Inglés

José Luis Leyva

Idea Editorial – www.ideaeditorial.com

Translapro Translator and Interpreter Network - www.translapro.com

Series: 1333 Most Frequently Used Terms

ISBN: 1490949240
ISBN-13: 978-1490949246

PREFACE

The purpose of this book is not only to serve as an English-Spanish reference work to look up a term when needed, but also as a guide to learn the most frequently used electrical terms. Learn just a few terms every day, and soon you will be acquainted with the most common electrical terminology in English and Spanish.

PREFACIO

El propósito de esta publicación –aparte de servir como obra de referencia donde se puedan consultar términos eléctricos cuando sea necesario- es poner al alcance del lector una sencilla guía con la que pueda familiarizarse con los términos eléctricos que más frecuentemente se utilizan. Si se aprenden tres o cuatro términos cada día, en poco tiempo aprenderá los términos eléctricos más comúnmente utilizados en inglés y español.

ACKNOWLEDGMENTS

During more than 25 years of work as an interpreter and translator, I have been in contact with people from many different areas of expertise. These people have provided me with the knowledge and wisdom from their technical fields. CEO's, manufacturing directors, human resources managers, plant managers, attorneys, ambassadors and even Presidents have in a way helped in creating the knowledge base for this book. Special thanks to Salvador Tarín and Francisco Leyva Álvarez, who -even inadvertently- contributed in this book.

RECONOCIMIENTOS

Durante los más de 25 años de mi profesión como intérprete y traductor, he conocido innumerables personas con diferentes contextos profesionales. Todas estas personas me han brindado el conocimiento y sabiduría correspondientes a su área del saber. Directores ejecutivos, directores de manufactura, gerentes de recursos humanos, gerentes de planta, abogados, embajadores y hasta presidentes, todos ellos de alguna manera han ayudado a integrar la base de conocimiento necesaria para crear esta publicación. Vaya un agradecimiento especial para Salvador Tarín y Francisco Leyva Álvarez, quienes —aun sin darse cuenta de ello- han contribuido en la realización de este libro.

ENGLISH-SPANISH
INGLÉS-ESPAÑOL

A

able to locate, fácil de localizar

above ground, sobre el nivel del suelo

abrasion, abrasión

AC circuit, circuito de CA

AC polyphase motor, motor polifásico CA

AC squirrel cage motor, motor en jaula de ardilla de CA

accessible, accesible

accurate, preciso

accurate reading, lectura precisa

activation of emergency voice communications, activación de comunicaciones de voz de emergencia

activation of the initiating device, activación del dispositivo de inicio

active component, componente activo

ADA handicapped electrical installation requirements, requisitos del ADA (Americans with Disabilities Act) para instalaciones eléctricas

adjacent structure, estructura adyacente

adjustment factor, factor de ajuste

administration, Administración

air conditioning compressor, compresor de aire acondicionado

air motor, motor neumático

air space, espacio aéreo

alkali chemicals, químicos alcalinos

alpha-numeric code, código alfanumérico

alternate current (AC), corriente alterna (CA)

alternate energy, fuente de energía alterna

alternate source of power, fuente alterna de corriente

alternating, alterna

alternating induced current, corriente alterna inducida

alternative energy, energía alternativa

aluminum, aluminio

aluminum conductor, conductor de aluminio

ambient sound level, nivel de sonido ambiental

ambient temperature, temperatura ambiente

American Wire Gauge Standards, Normas de Calibración Norteamericanas

ammeter, amperímetro

amorphous sealed device, dispositivo sellado amorfo

ampacity, amperaje

ampere, amperio

ampere reader ammeter, dispositivo para leer los amperios

amp-hour, amperio-hora

amps, amps

angle grinder, afiladora angular

annunciator, anunciador

antenna, antena

antenna system, sistema de antena

anti-oxidant paste, pasta antioxidante

arc chute, canal de descarga de arco

arc fault breaker, disyuntor de falla de arco

ARC fault circuit interrupter, interruptor de circuito de falla ARC

arc welding, soldadura de arco

arcing, arco

argon, argón

armored cable, cable blindado

attachment plug, clavija de conexión

attachment plug cap, tapa del tomacorriente

attachment point of a service drop, punto de conexión de un ramal de acometida

attic, entretecho

audible alarm, alarma sonora

audible appliances, dispositivos audibles

audio continuity tester, probador de continuidad de audio

automatic fire sprinkler system, sistema automático de aspersión contra incendios

automatic light-actuated device, dispositivo automático activado por la luz

auto-ranging, rango de entrada automático

available interrupted capacity, capacidad interrumpida disponible

average inrush capacity, capacidad de irrupción promedio

AWG, AWG

B

back-up time, tiempo de respaldo/reserva

ballast, balasta

banking, entibamiento

bare, sin aislamiento

bare conductor connection, conexión de conductos desnudos

bare copper conductor, conductor de cobre desnudo

barrier plate, placa de barrera

base plate, placa de soporte

base speed, velocidad de base

battery, batería

battery charger, cargador de baterías

battery leads, terminales de la batería

beam, viga

beam clamps, abrazaderas de haz

bell box, caja en campana

bend, doblar/curvar

bends, codos

beveler, biseladora

bipolar junction transistor, transistor de unión bipolar

blade contacts, contactos de hoja

blow, disparar

blow a fuse, fundirse un fusible

board, panel

bond wire, cable de empalme

bonded, empalmado

bonding, empalme

bonding jumper, puente principal

box fill, ocupación de una caja

box opening, abertura de una caja

box without a splice or tap, caja sin empalme o derivación

boxes in concealed work, cajas en obras ocultas

B-phase conductor, conductor de fase B

bracket, ménsula/soporte

branch circuit, circuito de derivación

branch circuit conductors, conductores de derivación

branch/shunt, derivación

brass junction box, caja de conexiones de bronce

breadboard, circuito experimental

breaker, disyuntor

buffer, regulador

build a circuit, armar un circuito

burial, instalación bajo tierra

buried cables, cables enterrados

burnish, bruñir

busbar, barra conductora

bushing, pasamuro

busway, canaleta de tránsito

butt splice connectors, conectores de empalme a tope

C

cabinet, gabinete

cable ampacity, amperaje de los cables

cable armor, blindaje de cables

cable clamp, pinza de cables

cable jacket, envoltura del cable

cable puller, tiracables

cable pulling rope, soga tiracables

cable sheath, cubierta para cables

cable tray, bandeja portacables

cable tray installation, instalación de bandeja portacables

cadweld mold, molde de "cadweld"

California style receptacles, tomacorrientes estilo "California"

canopy, cúpula

capacitance, capacitancia

capacitive, capacitivo

capacitor, condensador

capturing board, tarjeta de captura

carnival ride lighting, iluminación de carnaval

cartridge, cartucho

caution tags, etiquetas de precaución

ceiling, cielorraso

ceiling grids, rejillas de techo

cellular floor, piso de tipo celular

center to center distance, distancia de centro a centro

center to center measurement, medida de centro a centro

chain wrench, llave de cadenas

chamber, cámara

channel locks, bloqueos de canal

charge, carga

check, verificar

check list, lista de verificación

chemical reaction, reacción química

circuit, circuito

circuit board, tarjeta de circuitos

circuit breakers, cortacircuitos

circuit conductor, conductor del circuito

circuit interface, interfaz de circuitos

circuit voltage, voltaje del circuito

circular mil, milipulgada circular

circumference, circunferencia

circumference of the conduit, circunferencia del conducto

citrus plate, placa "citrus"

clamp-on ammeter, amperímetro de pinza

coax, coaxial

coaxial cable, cable coaxial

coil, bobina

coils of wire, bobina de alambre

cold chisel, cincel

combination circuit, circuito combinatorio

combined cross-sectional area of the wires, sección transversal combinada de los cables

combined voltage drop, caída de tensión combinada

combustible and non-combustible walls, paredes combustibles y no combustibles

combustible material, material combustible

combustion particles in the air, partículas de combustión en el aire

communication supply drops, bajadas de suministro de comunicación

communications circuit, circuito de comunicaciones

compact conductor connection, conexión de conductos compactos

compensate, compensar

component, componente

computed load, carga calculada

computer numerical control, control numérico por computadora

concentric bend, codo concéntrico

concrete block, bloque de concreto

concrete-encased electrode, electrodo empotrado en concreto

condenser, condensador

conducting body, cuerpo conductor

conductive part, parte conductora

conductive path, trazado conductor

conductivity, conductividad

conductor, conductor

conduit brush, cepillo para conductos

conduit die, troquelador de conductos

conduit routing, trayecto del conducto

conduit run, recorrido de un conducto

conduit runs, tendidos de los conductos

conduit swab, hisopo para conductos

connection to a raceway, conexión a una canaleta

connector, conector

consequent pole, polo consecutivo

constant rate descent, descenso a velocidad constante

constant-wattage - auto-transformer, autotransformador de vataje constante

continuous, continuo

continuous flow, flujo continuo

continuous load, carga continua

cooling system, sistema de enfriamiento

copper, cobre

copper alloy, aleación de cobre

copper conductor, conductor de cobre

copper-clad aluminum, aluminio revestido con cobre

copper-sheathed cable, cable con vaina de cobre

core, núcleo

corroded, oxidado

corroded metal, metal oxidado

corrosion, corrosión

corrosion protection, protección anticorrosiva

corrosive locations, lugares corrosivos

corrosive vapors, vapores corrosivos

corrugated metal, metal corrugado

coupling, acoplamiento

cover, cubierta

cover plate, cubierta protectora

cross sectional area, área transversal

crossed, intercambiados

crossed wires, alambres cruzados

cross-sectional area, área de la sección transversal

cross-sectional measurement, medición transversal

cross-sectional view, vista transversal

crystal, cristal

cubic inch, pulgada cúbica

current, corriente

current carrying ampacity, amperaje de transporte de corriente

current to fluctuate, fluctuaciones de corriente

current-carrying capacity, capacidad de transporte de corriente

current-transformer principle, principio de transformador de corriente

cut end, extremo de corte

cutting torch, soplete de corte

D

data processing unit, unidad de procesamiento de datos

DC motor, motor de corriente directa

DC voltage, voltaje/tensión de CD

de-burr, desbarbar

decibels, decibeles

deck, plataforma/cubierta

deduction adjustment, ajuste por reducción

degree of bend, grado de curvatura

degrees, grados

de-icing equipment, equipo de descongelamiento

demagnetize, desmagnetizar

demand factor, factor de demanda

de-oxide compound, compuesto desoxidado

depression, depresión

depth, profundidad

derating, disminución de potencia

designation EOL, indicación EOL

destructive conditions, condiciones destructivas

device, dispositivo

device box, caja de dispositivos

diagnose, diagnosticar

diagnosis, diagnóstico

diagnostic trouble code, código de falla del diagnóstico

diagram, diagrama

dial indicator, indicador de reloj

diameter, diámetro

diameter of hole, diámetro del orificio

die grinder, afiladora recta

difference, diferencia

differential temperature, temperatura diferencial

digital logic, lógica digital

digital logic fundamentals, fundamentos de lógica digital

digital multimeter, multímetro digital

digital signal processor (DSPS), procesador digital de señales

digital signals, señales digitales

dim, atenuar

dimension, dimensión

dimming, reductor

diode, diodo

direct current (DC), corriente directa (CD)

directly proportional, directamente proporcional

disassemble, desmontar

disconnecting means, medios de desconexión

disk-aimer, orientador de antena

distance, distancia

distribution, distribución

double lug, terminal de orejeta doble

drag, arrastre

drill bit, broca

drip loops, lazos de goteo

drip proof, a prueba de salpicaduras

drive axle, eje motor

drive circuit, circuito excitador

driven ground rod, varilla de tierra conducida

driven machinery, maquinaria impulsada

drop, caída

dry type transformer, transformador de tipo seco

dual element fuse, fusible de dos elementos

dual pushbutton switch, interruptor de botón pulsador doble

dual trace oscilloscope, osciloscopio de doble traza

duo-servo, doble servo

dust-free, libre de polvo

dust-ignition proof, a prueba de ignición-polvo

dustproof, a prueba de polvo

E

earth returns, retornos a tierra

edge, borde

electric arc, arco eléctrico

electric hedge trimmer, tijera eléctrica

electric motor, motor eléctrico

electric relay, relé/relevador

electric shock, choque eléctrico

electrical circuit, circuito eléctrico

electrical contact, contacto eléctrico

electrical continuity, continuidad eléctrica

electrical current, corriente eléctrica

electrical device, dispositivo eléctrico

electrical engineer, ingeniero eléctrico

electrical equipment, equipo eléctrico

electrical failure, falla eléctrica

electrical fire, incendio eléctrico

electrical interlock, interbloqueo eléctrico

electrical load, carga eléctrica

electrical metal tubing, tubería eléctrica de metal

electrical metallic tubing, cañería eléctrica metálica

electrical metallic tubing (EMT), tubería eléctrica metálica (EMT)

electrical noise, ruido eléctrico

electrical nonmetallic (Smurf Tube) tubing, tuberías eléctricas no metálicas (tubo Smurf)

electrical outlet box, caja de salida eléctrica

electrical penetration, penetración eléctrica

electrical power, energía eléctrica

electrical power supply, fuente de energía eléctrica

electrical rigid metal conduit, conducto eléctrico de metal rígido

electrical room, sala eléctrica

electrical rough-in, boceto eléctrico

electrical shock, electrochoque

electrical signals, señales eléctricas

electrical space, espacio eléctrico

electrical wall switch, interruptor eléctrico de pared

electrical warnings, advertencias eléctricas

electrical wiring, cableado eléctrico

electrician, electricista

electrician's tape, cinta de electricista

electricity, electricidad

electrolyte, electrólito

electromagnet, electroimán

electromagnetic device, dispositivo electromagnético

electromagnetic interference, interferencia electromagnética

electromagnetic wave, onda electromagnética

electromagnetism, electromagnetismo

electromechanical, electromecánico

electromotive force, fuerza electromotriz

electron beam, haz de electrones

electronic ignition, encendido electrónico

electronics technician, técnico en electrónica

electrostatic, electrostático

elongation, alargamiento

emergency lighting, luces de emergencia

emergency lighting fixtures, accesorios para iluminación de emergencia

emergency power, energía de emergencia

emergency power source, fuente de energía de emergencia

emergency standby generator, generador de reserva de emergencia

emergency transfer switch, interruptor de transferencia de emergencia

emitter of light, emisor de luz

EMT, EMT (tuberías metálicas eléctricas)

encase, empotrar

encircle, rodear

enclosure, recinto

end, extremo

endothermic reaction, reacción endotérmica

energy alternate sources, fuente alterna de energía

enforcement, cumplimiento

envelope, revestimiento

equipment, equipo

ER fitting, accesorio ER

exit lighting fixtures, artefactos de luces de salida

exothermic (cadwelding), exotérmico (soldadura de cadmio)

exothermic reaction, reacción exotérmica

exothermic welding, soldadura exotérmica

expandable conduit, conducto expansible

expansion joint, empalme de expansión

expansion screw anchors, anclajes de tornillos de expansión

expansive soils, suelos expansivos

explosion-proof, a prueba de explosión

exponential, exponencial

exposed to oil and gasoline, expuestos al aceite y la gasolina

exterior and interior metallic finish, acabado metálico exterior e interior

exterior manual switches, interruptores manuales exteriores

extra cable, cable adicional

extruded polymers, polímeros extruidos

F

failure, falla

farads, faradios

fault, falla

fault current, corriente de falla

fault insertion switch, interruptor de inserción de fallas

feed roller, rodillo alimentador

feeder, alimentador

feeder ampacity, ampacidad del alimentador

feeder and branch circuit, circuito de alimentación y derivación

feeder circuit, circuito de alimentación

feeder load, carga del alimentador

feeder tap rule, regla de la toma del alimentador

feet, pies

ferrous raceway, canaleta ferrosas

fertilizer, fertilizante

festoon, realce

festoon lighting, iluminación de realce

fiber, fibra

fiber optics, fibra óptica

fiberglass, fibra de vidrio

fiber-reinforced bakelite, bakelita reforzada con fibra

field bend, codo de campo

field line, línea de campo

field-effect transistor, transistor de efecto de campo

filament, filamentos por pulgada

fire alarm device, dispositivo de alarma contra incendio

fire pump, bomba contra incendio

fire stopper, cortafuego

fire suppression system, sistema de supresión de incendios

fire-retardant material, material retardador de incendios

fish tape, alambre guía

fixed trip, disparo fijo

fixed wiring method, método de cableado fijo

fixture ballast, balasto de accesorio

fixture stud, porta-accesorios

fixture terminal, terminal del accesorio

flame detector, detector de llama

flame-retardant, retardador de llama

flammable, inflamable

flammable material, material inflamable

flashing red strobe light, luz estroboscópica roja titilante

flexible bonding jumper, puente de conexión flexible

flexible cable, cable flexible

flexible connection, conexión flexible

flexible cord, extensión flexible

flexible hose clamp, abrazadera para manguera flexible

flexible metal conduit, conducto metálico flexible

float switch, interruptor de flotador

floating voltage system, sistema de voltaje/tensión flotante

flow control, control de flujo

flowmeter, flujómetro

fluctuating voltage, fluctuación de voltaje/tensión

fluorescent, fluorescente

fluorescent lamp, lámpara fluorescente

flush raceway, canaleta embutida

flush with the finished face of the wall, al ras de la superficie terminada de la pared

footing, zapata

for portable use only, sólo para uso portátil

force of magnetism, fuerza magnética

forward motion, movimiento hacia delante

four-wire Delta system, sistema Delta de cuatro cables

FPL cable, cable PFL

FPLP type cable, cable de tipo FPLP

frequency, frecuencia

friction, fricción

full-load current, corriente a plena carga

fuse, fusible

fuse box, caja de fusibles

fuse holder, portafusibles

G

gallon, galón

galvanized conduit, conducto galvanizado

galvanized steel, acero galvanizado

gap, brecha/hueco

gas welding, soldadura de gas

gas-filled chambers, cámaras con circulación de gas

gauge, calibre

general wiring, cableado general

generator, generador

generator set, conjunto de generador

glue, pegamento

gooseneck, cuello de cisne

gravity operated, gravitacional

grip, apriete

grommet, arandela aislante

ground currents, corrientes de tierra

ground fault, falla de tierra

ground fault circuit interrupter (GFCI), interruptor accionado por corriente de pérdida a tierra

ground fault condition, condición de falla a tierra

ground fault protection, protección contra falla a tierra

ground ring, anillo de masa

ground rod, varilla de puesta a tierra

ground wire, cable a tierra

grounded circuit conductor, conductor del circuito conectado a tierra

grounded conductor insulation, aislamiento del conector puesto a tierra

grounded electrode conductor, conductor del electrodo conectado a tierra

grounded metallic covers, cubiertas metálicas puestas a tierra

grounded metallic plates, placas metálicas puestas a tierra

grounded service conductor, conductor de servicio conectado a tierra

grounded system, sistema conectado a tierra

ground-free, libre de tierra

grounding clip, presilla de puesta a tierra

grounding conductor, conductor de puesta a tierra

grounding electrode, electrodo de puesta a tierra

grounding electrode conductor, conductor del electrodo de puesta a tierra

grounding extension cords, cables de tipo a tierra

grounding portable generator, generador de puesta a tierra portátil

grounding requirements, requisitos de conexión a tierra

grounding type, tipo apto para conexión a tierra

ground-start, tierra-arranque

gutter, canalón

H

hacksaw, sierra para metales

halogen iode distribution, distribución de yodo halógeno

halogen tubular lamp, lámpara tubular de halógeno

hammer, martillo

hand bender, dobladora manual

hand tool, herramienta manual

handhole, agujero de inspección

handle, asa

hanging, pendiente

hard automation, automatización dura

hardness, dureza

harmonic, armónico

heat, calor

heat energy, energía calórica

heat insulator, aislador de calor

heat of the arc, calor del arco

heat resistant, resistente al calor

heat sensor, sensor de calor

heat sink, aislador térmico

heater, calentador

heat-shrink, termocontracción

heat-shrink insulators, aisladores termoencogibles

heavy industrial distribution, distribución industrial pesada

height, altura

heliarc ion discharge, descarga de ión de arco de helio

helium, helio

hertz (Hz), hertzios (Hz)

hex key, llave hexagonal

HID, HID

high, alto

high intensity discharge, descarga de alta intensidad

high pressure sodium (Lucalox), sodio de alta presión (Lucalox)

high voltage, alta tensión/alto voltaje

high voltage welding current, corriente de soldadura de alta tensión

high-load, carga alta

highly compressed refractory insulation, aislamiento refractario altamente comprimido

hinged base, base abisagrada

hole, orificio

hollow space, espacio hueco

horizontal elbow, codo horizontal

horizontal splice plates, placas de empalme horizontal

horizontal Tee, T horizontal

horn strobe, señal de bocina del estrobo

horsepower, caballo de fuerza

horsepower rating, caballos de fuerza nominales

hot and neutral conductors, conductores vivo y neutro

hot phase conductor, conductor de fase caliente

hot water circulating pump, bomba de circulación de agua caliente

hot wire to the ground, cable con tensión a tierra

housing, cubierta protectora

hydraulic bender, dobladora hidráulica

hydraulic press, prensa hidráulica

hydraulic tool, herramienta hidráulica

hydrogen gas, gas hidrógeno

I

identified, identificado

ignition point, punto de ignición

illuminate, iluminar

impedance, impedancia

in parallel, en paralelo

in series, en serie

incandescent fixture, dispositivo incandescente

inches, pulgadas

increase, aumentar

incremental positioning, posicionamiento incremental

independent support wires, alambres de soporte independientes

indicator, indicador

inductance, inductancia

induction, inducción

induction motor, motor de inducción

inductive, inductivo

inductive current, corriente inductiva

inductor, inductor

information technology system, sistema de tecnología de la información

infrared radiation, radiación infrarroja

initiation device, dispositivo de iniciación

input, entrada

input distance, distancia inicial

input force, fuerza aplicada

input to the motor, entrada al motor

inside diameter, diámetro interior

inspector's test valve, válvula de prueba de inspección

installation, instalación

installation with multiple services, instalación con múltiples servicios

instrument scan, verificación por instrumentos

instrumentation system, sistema de instrumentación

insulated, aislado

insulated bushing, buje aislado

insulated conductor, conductor aislado

insulated step, peldaño aislado

insulating material, material aislante

insulation, aislamiento

insulation resistance, resistencia de aislamiento

insulator, aislador

integrated circuit, circuito integrado

integrated electrical system, sistema eléctrico integrado

interchangeable, intercambiable

interconnected electrical power production, sistema de producción de energía eléctrica

interior hollow masonry partition, tabique de mampostería huecos

interrupt, cortar

interrupting rating, capacidad de corte

interval, intervalo

inverse time circuit breaker, disyuntor de tiempo inverso

inversely proportional, inversamente proporcional

iron core, núcleo de hierro

irreversible compression-type connector, conector de tipo de compresión irreversible

irrigation pump, bomba de irrigación

isolated circuit, circuito aislado

isolated ground type receptacles, tomacorriente de tipo de tierra aislado

isolating switch, interruptor aislador

isometric, isométrico

J

janitorial equipment, equipo de conserjería

joists, cabriada

jumper cable, cable de conexión

junction box, caja de conexiones

K

Kellum's Grip, mordaza de Kellum

keyhole saw, sierra de calar

kick, reductor

kilovolts-amperes (kVA), kilovoltios-amperios (kVA)

kilowatt-hour, kilovatio-hora

kilowatts (kW), kilovatios (kW)

knife disconnect, disyuntor de cuchilla

L

ladder, escalera

ladder logic, lógica en escalera

lag shield, manguito de expansión

lamp, lámpara

lampholder, portalámpara

landscape lighting, iluminación paisajística

lead, plomo

lead acid battery, batería de plomo-ácido

life-safety branch, rama de seguridad de vida

lifting hook, gancho para levantar

light bulb, bombilla eléctrica

light emitting diode (LED), diodo emisor de luz (LED)

light rail system, sistema de riel de iluminación

lighting load, carga de iluminación

lightning arrester, pararrayos

limit switch, interruptor limitador

line, línea

line voltage, voltaje/tensión de línea

linear measurement, medición lineal

linear motion, movimiento rectilíneo

linearity, linealidad

liquid tight, impermeable a los líquidos

listed lug, orejeta reglamentaria

live part, parte con voltaje/tensión

load, carga

load conductor connection, conexión de conductos de carga

load current, corriente de carga

load factor, factor de carga

load served, carga servida

load voltage, voltaje/tensión de carga

loadstone, magnetita

local metallic underground system, sistema subterráneo metálico local

lock, bloquear

lock condition, condición cerrada

locked rotor amperes, amperios del rotor enclavado

locked-rotor current, corriente del rotor enclavado

locking type, tipo de fijación

locknut, contratuerca

lockout/tagout, bloqueo/etiquetado de energía

logic gate, acceso de lógica

long-life battery, batería de larga vida

looped, en bucle

loop-start, bucle-arranque

loose connection, conexión floja

low battery power, baja energía en las baterías

lubricating gel, gel lubricante

lumen, lumen

M

machine-ground matching surface, superficie de adaptación de máquina-tierra

magnet, imán

magnetic cutout, cortacircuito magnético

magnetic field, campo magnético

magnetic starter, arrancador magnético

magnetic strength, fuerza magnética

magnetic tool, herramienta magnética

magnetite, magnetita

magnetization, imantación

main bonding jumper, puente principal

main power indicator, indicador principal de energía

mandatory, obligatorio

manometer, manómetro

manual pull box, caja de acceso manual

masonry, mampostería/albañilería

masonry block, bloques de mampostería

matrix, matriz

maximum charge, carga máxima

maximum distance, distancia máxima

measure of motor efficiency, medida de la eficiencia del motor

measuring tape, cinta de medir

mechanical plan, plano mecánico

mechanical press, prensa mecánica

megawatts (MW), megavatios (MW)

megger, megaóhmetro

megohmmeter, megaóhmetro

melting point, punto de fusión

messenger wire, cable portador

metal conduit, conducto metálico

metal cutting tool, herramienta para corte de metal

metal frame of the structure, marco metálico de la estructura

metal halide, halogenuro metálico

metal pole, poste metálico

metal raceway, canaleta de metal

metal underground gas piping, tubería de gas subterránea

meter, medidor/metro

meter malfunction, mal funcionamiento del medidor

meter-in, medidor de entrada

metering device, dispositivo de medición

meter-out, medidor de salida

micrometer, micrómetro

microprocessor applications board, placa de aplicaciones del microprocesador

microslide, microplatina

microslide viewer, visor microscópico

milliameter, miliamperímetro

milliamp signal system, sistema de señal de miliamperios

milliamps, miliamperios (ma)

mineral insulated cable (MI), cable mineral aislado (MI)

Minerallac support, soporte Minerallac

minimum, mínimo

minimum distance, distancia mínima

modulated laser, láser modulado

moisture, humedad

moisture-resistant thermoplastic, termoplástico resistente a la humedad

molded-case circuit breaker, disyuntor de circuito en caja moldeada

molly-or toggle-type fasteners, fijadores atravesados ("molly")

momentary load, carga momentánea

motor branch circuit, circuito derivado de un motor

motor control, control regulador de motor

motor control center, centro de control de motores

motor controller, controlador de motor

motor frame, marco del motor

motor lead wire, hilo conductor del motor

motor mounting plate, placa de montaje del motor

motor nameplate, rótulo del motor

motor operated appliance, dispositivo que funciona a motor

motor pushbutton, pulsador del motor

motor run capacitor, capacitor para funcionamiento de motores

motor specifications, especificaciones del motor

motor starter, arrancador de motor

motor terminal module, módulo terminal del motor

mount, montar

mounting strap, correa de instalación

movable contact, contacto móvil

multi-branch circuit, circuito de múltiples derivaciones

multiconductor cable, cable multiconductor

multiconductor ribbon cable, cable plano de múltiples conductores

multifamily unit, unidad multifamiliar

multimeter (VOM), multímetro (VOM)

multi-outlet assembly, conjunto de múltiples salidas

multi-tap, toma múltiple

multi-wire branch, bifurcación con cables múltiples

multiwire circuit, circuito multicableado

Myers hub, concentrador Myers

N

nameplate, rótulo

national electrical code, código eléctrico nacional

needle, aguja

needle nose pliers, pinzas de punta de aguja

needle scaler, martillo de agujas

negative, negativo

negative pole to ground, del polo negativo a tierra

negative pole to positive pole, del polo negativo al positivo

NEMA enclosure, caja NEMA

NEMA rating, régimen NEMA

NEMA Type 3R motor controller, controlador de motor tipo 3R NEMA

neon, neón

neon lighting, luces de neón

net, malla

network, red

network communications, comunicaciones en red

neutral conductor insulation, aislamiento del conductor neutro

neutral safety switch, interruptor neutral de seguridad

neutral wire, cable neutro

niche, nicho

nickel, níquel

nickel-clad copper, cobre revestido en níquel

nipple, niple

no equipment ground, ningún equipo puesto a tierra

nominal voltage, voltaje/tensión nominal

non-combustible cover, cubierta no combustible

nonconductive coatings, revestimiento no conductor

nonincendive field wiring, cableado de campo ignífugo

nonmetallic cable, cable no metálico

nonmetallic conduit, conducto no metálico

nonmetallic cover, cubierta no metálica

nonmetallic sheathed cable, cable con vaina no metálica

non-metallic sheathed cable, cables con cubierta no metálica

non-power limited, sin limitación de potencia

non-power limited circuit, circuito sin potencia limitada

non-skid feet, peldaños antideslizantes

notch, entalladura

number of pounds per foot of wire, número de libras por pie de cable

number of threads per inch, número de roscas por pulgada

numeric keypad, teclado numérico

nut, tuerca

O

obstruction, obstrucción

occupational safety and health, seguridad y salud en el trabajo

octagon outlet box, caja octagonal

offset, desviación

ohm, ohm

ohmmeter, ohmímetro

ohm's law equation, ecuación de la ley de ohm

oil-filled chamber, cámara con circulación de aceite

oil-filled device, dispositivo lleno de aceite

oil-filled transformer, transformador con circulación de aceite

on time delay, tiempo de retardo

one-hole strap, correa de un orificio

one-phase, monofásico

open, abierto

open circuit, circuito abierto

open circuit voltage, voltaje del circuito abierto

open ground, interrupción a tierra

operating current, corriente de funcionamiento

operation, operación

operational, operativo

optical fiber, fibra óptica

optical line tester, comprobador de línea óptico

opto-coupler, opto acoplador

original elevation, elevación original

oscillator, oscilador

oscilloscope, osciloscopio

outdoors, exterior

outer sheath of copper or steel, vaina externa de cobre o acero

outgoing and return conductor, conductor de salida y retorno

outlet, tomacorriente

outlet box, caja de salida

outline, contorno

output, salida

output distance, distancia resultante

output of the motor, salida del motor

outside diameter, diámetro externo

overcurrent, sobrecorriente

overcurrent device, dispositivo de sobrecorriente

overcurrent protection, protección contra sobrecorriente

overhead drop, bajada general

overhead feeder, alimentador aéreo

overheat, sobrecalentar

overheating, sobrecalentamiento

overheating under load, sobrecalentamiento bajo carga

overload, sobrecarga

overload current, corriente de sobrecarga

overload heater, calentador de sobrecarga

overload protection, protección contra sobrecarga

overload relay, relé de sobrecarga

overvoltage, sobretensión

P

packaged assembly, conjunto integrado

panelboard, panel de control

parallax, paralaje

parallel, paralelo

parallel circuit, circuito en paralelo

partition, tabiques

party-line, compartido

passive component, componente pasivo

passive equalizer system, sistema ecualizador pasivo

path, ruta

pattern, patrón

peak voltage, voltaje máximo

peen hammer, martillo de bola

pendant-type fluorescent fixture, dispositivo fluorescente colgante

percent, porcentaje

percent slip, porcentaje de deslizamiento

period, período

phase converter, conversor de fase

phase meter, fasímetro

phase synchronization, sincronización de fases

photo-voltaic, fotovoltaico

Pig tail, acoplamiento metálico flexible "Pig Tail"

pig tailed, enrollado en espiral

pipe cutter, cortacaños

pipe die, troquelador de caños

pipe grease, grasa de cañerías

pit, pozo

pitch, cuerda

pivot, pivote

plaster ring, anillo de yeso

plastic hammer, martillo de plástico

plate of iron or steel, placa de hierro o acero

platform, plataforma

PLC, controlador lógico programable

plenum, cámara de distribución

plug, clavija

plug and receptacle combination, combinación de clavija y tomacorriente

plug connector, conector de clavija

plug set, conjunto de clavijas

plumb-bob, plomada

plywood, triplay

pneumatic hoist, polipasto neumático

pneumatic press, prensa neumática

pneumatic tool, herramienta neumática

point of attachment, punto de conexión

point to point wiring diagram, diagrama de cableado entre puntos

pointer, aguja indicadora

polarity, polaridad

polarity-sensitive device, dispositivo sensible a la polaridad

polyurethane, poliuretano

polyvinyl chloride, cloruro polivinílico

portable, portátil

positive, positivo

positive pole to ground, del polo positivo a tierra

positive to negative pole, del polo positivo al negativo

potential transformer, transformador de potencial

potentiometer, potenciómetro

pothead, cabeza terminal de cable

pounds, libras

poured concrete, concreto vaciado

power, potencia

power factor, factor de potencia

power mosfet, mosfet de potencia

power source, fuente de energía

power supply, suministro de energía

power thyristor, tiristor de potencia

power transmission, transmisión de energía

power-limited, limitación de potencia

preliminary check, prueba preliminar

press, prensa

pressed board, cartón prensado

pressure switch, interruptor de presión

pressurized water tank, tanque de agua presurizado

pretest, preevaluación

pre-wired, precableado

primary coil, bobina primaria

printed circuit board, tarjeta de circuitos impresos

programmable logic controller, controlador lógico programable

propane torch, soplete de propano

proton particles, partículas de protones

proximity, aproximación

psi, psi

pull point, punto de acceso

pull station, puesto de tracción

pulley, polea

pulling basket, cesta de arrastre

pulling compound, compuesto de arrastre

pulsating, pulsante

pump, bomba

PVC conduit, conducto de PVC

PVC coupling, acoplamiento de PVC

Q

quartz, cuarzo

R

raceway, canaleta

radiation, radiación

radio broadcasting, radiodifusión

radius bend, codo radial

radius of the wires, radio de los cables

raintight, impermeable a la lluvia

raintight seal, sello a prueba de lluvia

rate of rise detector, detector de velocidad de aumento

rated ampacity, amperaje nominal

rated current, corriente nominal

rated horsepower, régimen de caballos de fuerza

rating, régimen

rating of the bussing, régimen del conductor común

Rawl-drives, pernos Rawl

RC time constant, constante de tiempo de circuito RC

reacceptance test, prueba de reaceptación

reactor, reactor

read and write cycle, ciclo de lectura y escritura

readability, legibilidad

ream, escariar

receiving equipment, equipo receptor

receptacle, receptáculo

receptacle enclosure, recinto para tomacorriente

recessed, sobresaliente

recessed incandescent fixture, accesorio incandescente empotrado

rectified alternating current, corriente alterna rectificada

rectifier, rectificador

reel, carrete

reel jack, porta-carretes

reflected ceiling plan, plano de techo reflejado

refrigeration system, sistema de refrigeración

regulate speed, regular la velocidad

regulator, regulador

reinstall, reinstalar

relay, relé/relevador

release of toxic fumes, emisión de humos tóxicos

relocatable wired partitions, tabiques cableados reubicables

remote supervising station, estación de supervisión remota

remote-control circuit, circuito de control remoto

removable, removible

removable ladder, escalera removible

removing card, tarjeta de extracción

renewable energy, energía renovable

replace, reemplazar

replacement, sustituto

replacement link, cinta de recambio

reprogram, reprogramar

requirements of the spacing, requisitos de espaciado

reset, reposición

resettable fuse, fusible restaurable

residential branch circuit, circuito derivado residencial

resistance, resistencia

resistance factor, factor de resistencia

resistive load, carga resistiva

resistivity scale, escala de resistividad

resistor, resistor

resonant frequency, frecuencia resonante

reverse polarity, polaridad inversa

rheostat, reóstato

rigid IMC and EMT conduit, conducto IMC y EMT rígido

rigid metal conduit, conducto metálico rígido

riser diagram, diagrama ascendente

riveting hammer, martillo remachador

robotic arm, brazo robot

rod, varilla

rods of nonferrous materials, varillas de materiales no ferrosos

roller, rodillo

Romex cable system, sistemas de cables Romex

Romex connector, conector Romex

rotary switch, conmutador rotativo

rotor construction, construcción de rotor

rough edge, borde burdo

rubber washer, arandela de goma

rubber-insulated barrel connector, conductor con barril aislado con goma

running current, corriente de marcha

rusted part, parte oxidada

rusting, oxidación

S

saddle bend, codo en caballete

safety, seguridad

safety chain, cadena de seguridad

safety latch, cierre de seguridad

safety regulations, normatividad de seguridad

safety-control equipment, equipo de control de seguridad

salient-pole, polo salientes

sand, lijar/arena

satellite dish, antena parabólica

scar, rasgar

schematic, esquema

schematic symbol, símbolo esquemático

screw, tornillo

screw shell, casco del tornillo

screwdriver, destornillador

sealing hub, cubo sellador

sectional drawing, trazado de secciones

self-acting, de acción automática

self-contained unit, unidad autónoma

self-monitoring device, dispositivo de auto-monitoreo

semi-automatic, semiautomático

semi-conductor, semiconductor

separate path, trazado separado

separately derived system, sistema derivado por separado

sequence of control operations, secuencia de operaciones de control

sequencing, secuencial

serial port, puerto en serie

series circuit, circuito en serie

series-parallel circuit, circuito mixto/combinado

service bonding jumper, puente de conexión a tierra de servicio

service drop, caída de servicio

service entrance conductor, conductor de la entrada de servicio

service head, acometida de servicio

service switch, interruptor de servicio

service-drop conductor, conductor del ramal de acometida

servicing, reparaciones

servomechanism, servo-mecanismo

set back, retranqueo

set screw, tornillo de fijación

setting frequency, regulación de frecuencia

setup, configuración

shallow chase in concrete, ranura superficial en concreto

sheet metal screw, tornillo autorroscante

sheet rock, tablarroca

shell, carcasa

shielded, blindado

shoring, apuntalamiento

shoring of trenches, apuntalamiento de zanjas

short circuit, cortocircuito

short circuit current, corriente de cortocircuito

short circuit protection, protección contra cortocircuitos

short-circuit/ground fault fuse, fusible de cortocircuito/falla a tierra

shortening, acortamiento

signal circuit, circuito de señales

signaling, señalización

signaling line circuit (SLC), circuito de línea de señalización

silicone rubber sealed joint, junta sellada con goma de silicón

silver screw, tornillo plateado

sine wave, onda sinusoidal

single line diagram, diagrama de línea única

single pair shielded, cable de un par blindado

single phase, monofásico

single-phase three-wire system, sistema trifilar monofásico

single-pole double-throw (SPDT) switch, interruptor bidireccional unipolar

single-throw knife switch, interruptor de cuchilla unidireccional

site plan, plano de ubicación

size, tamaño

skewed, oblicuo

sky hook, gancho aéreo

slide, platina

slideway, guía de deslizamiento

smog testing machine, máquina de análisis de humo

smoke compartment, compartimento con humo

smoke detector, detector de humo

smoke spread prevention, prevención de la diseminación de humo

solar photo-voltaic system, sistema fotovoltaico solar

solder, soldadura

soldered connection, conexión soldada

soldering iron, cautín

solenoid, solenoide

solenoid-operated, operado por solenoide

solid, sólido

solid-state, estado sólido

source voltage, voltaje primario

spacer, espaciador

span, recorrido

sparking, chispas

special tool, herramienta especial

specifications, especificaciones

specify, especificar

splice plate, placa de empalme

split bolt, perno dividido

split-bolt connector, conector de perno dividido

spray lubricant, lubricante en atomizador

spring nut, tuerca de resorte

sprinkler system controller, controlador del sistema de aspersión

square box, caja plana

square inches, pulgadas cuadradas

square meter, metro cuadrado

squirrel cage, jaula de ardilla

squirrel cage motor, motor en jaula de ardilla

stabilize, estabilizar

stall light indicator, indicador de luz de detención

standard, norma

standard conditions, condiciones normales

standard Edison-base, base Edison estándar

standby generator system, sistema de generador de reserva

standby system, sistema de reserva

staple, grapa

star and box, estrella y caja

start, puesta en funcionamiento

start and stop circuit current, activar y desactivar la corriente del circuito

start/stop station, estación de arranque/parada

starter, dispositivo de arranque

starting torque, par de torsión de arranque

static charge of electricity, carga estática de electricidad

stationary contact, contacto fijo

stationary motor, motor estacionario

statistical process control, control estadístico del proceso

steel reinforcing bar, barra de refuerzo de acero

step fuse, fusible de paso

step-down, rebajador

step-up or step-down of voltage, elevar o reducir la tensión

stereo optic, estereoscópico

storage areas, áreas de almacenamiento

storage battery, acumulador

storage battery rooms, salas de acumuladores

strain, tensión

stranded, trenzado

stranded conductor, conductor trenzado

strapping, colocación de flejes

stray current, corriente parásita

stress, esfuerzo

stress analysis, análisis de esfuerzos

string, secuencia

stroke, carrera/desplazamiento

structural ceiling, techo estructural

structural member, miembro estructural

structural steel, acero estructural

stub-up, chicote

sub-panel, subpanel

subscript, subíndice

super glue, super-pegamento

supply, suministro

supply conductors, conductores de suministro

supply station, estación de alimentación

support, soporte

support of fixtures, soporte de accesorios

surface in contact with exterior soil, superficie en contacto con el suelo exterior

surface tension, tensión superficial

surfactant, surfactante

surge, sobretensión transitoria

surge arrester, disipador de sobretensión

survivability, capacidad de supervivencia

sweep bend, codo de barrido

switch, interruptor

switch leg, hilo exterior del interruptor

switchboards, tableros de distribución

switchgear, equipo de distribución

switching or control devices, dispositivos de conmutación o control

swivel joint connection, conexión de empalme giratoria

swivel plate, placa giratoria

symbol legend, referencia a símbolos

synchro- servo, servo sincronizado

synchronous and delta, sincrónica y delta

system, sistema

T

tab, lengüeta

tag, etiqueta/identificación

talk mode, modo de conversación

tank, tanque

taper pin, pasador cónico

taper proof, a prueba de golpes

tapered reamer, escariador cónico

T-bar ceiling, techo con una barra en T

Tek screw, tornillo Tek

temperature detector, detector de temperatura

temperature resistant glass, vidrio resistente a la temperatura

temporary circuit, circuito temporal

terminal, terminal

terminal lugs, terminales de orejetas

terminal strip, tira de bornes

test equipment, equipo de prueba

test lead, cable de prueba

test point, punto de prueba

thermal contraction, contracción térmica

thermal insulation, aislamiento térmico

thermal overload, sobrecarga térmica

thermal protector, protector térmico

thermal shock, choque térmico

thermally protected, con protección térmica

thermal-with nylon, térmico con nylon

thermistor, termistor

thermocouple, termopar

thermoelectric, termoeléctrico

thermometer alarm, alarma de termómetro

thermoplastic, plástico térmico

thermoset-waterproof, termoestable a prueba de agua

THHN wire, cable THHN

thickness, grosor

thread, rosca

thread gauge, calibrador de roscas

threaded, roscado

threaded couplings, acoples roscados

threaded hub, cubo roscado

threaded joints, uniones roscadas

threadless coupling, acople sin rosca

threads per inch, roscas por pulgada

three-dimensional, tridimensional

thyristor, tiristor

TIC tracer, trazador TIC

tie wire, conexión alámbrica

tight, ajustar

time-delay, retraso

time-delay fuse, fusible de retardo

time-delay relay, relé de retardo de tiempo

tire rim, rueda de un neumático

toggle, palanca

tool holder, porta-herramientas

torque, torque

total ampere ratings of the units, régimen total de amperios de las unidades

total cross-sectional fill area, área transversal de ocupación total

total wattage of the lamps, vataje total de las lámparas

tracer robot, robot trazador

trade size, tamaño comercial

transducer, transductor

transfer, transferir

transfer equipment enclosure, recinto de los equipos de transferencia

transfer switch, interruptor de transferencia

transfer time, tiempo de transferencia

transformer, transformador

transformer, transformador

transformer loss, pérdida del transformador

transient, transitorio

transistor, transistor

transmission mechanism, mecanismo de transmisión

transportation, transporte

traslational motion, movimiento de traslación

trip unit, unidad de disparo

tri-state, tres-estados

troubleshooting information, información sobre detección de fallas

tugger, remolcador

tungsten halogen lamp, lámpara halógena de tungsteno

twinax, cable de conductores axiales retorcidos

twisted pair, par torcido

twisting, torsión

twist-lock, cierre por torsión

two dimensions, bidimensional

two post connector, conector de dos patas

type MC cable, cable tipo MC

type MI cable, cable tipo MI

type NM cable, cable de tipo NM

Type S fuse, fusible tipo S

type THHN conductor, conductor tipo THHN

U

UF cable, cable de UF

Ufer ground, puesta a tierra Ufer

U-ground contact, contacto de tierra en U

ultra violet signature, señal ultravioleta

ultra-violet resistant paint, pintura resistente a la luz ultravioleta

under load, sometido a carga

under voltage, bajo voltaje

underground, subterráneo

underground wiring, cableado subterráneo

underwater lighting, iluminación subacuática

ungrounded conductor, conductor no conectado a tierra

ungrounded system, sistema sin puesta a tierra

ungrounded type, tipo no apto para conexión a tierra

unheated conduit, conducto no caldeado

uninterruptible power supply, suministro de energía no interrumpible

Unistrut material, material Unistrut

universal type motor, motor de tipo universal

update, actualizar

V

variable, variable

variable resistor (V.R.), resistor variable (RV)

variable setting, configuración variable

varistor, varistor

varying frequency, frecuencia variable

vault, bóveda

vehicle-mounted generator, generador montado sobre vehículos

vertical elbow, codo vertical

vertical Tee, T vertical

vibration, vibración

vibration isolator, aislante de vibración

virtual instrumentation, instrumentos virtuales

visible notification appliance, dispositivo de notificación visible

volt, voltio

volt amperes, voltamperios

voltage, voltaje/tensión

voltage converter, convertidor de voltaje/tensión

voltage drop, caída de voltaje

voltage regulator, regulador de voltaje

voltage surge, subida de voltaje

voltage tester, analizador de tensión

voltage to ground, voltaje a tierra

voltage under loaded conditions, voltaje en condiciones de carga

volt-amps, voltamperios

voltmeter, voltímetro

W

wall plug box, caja de clavijas de pared

water level switch, interruptor a nivel del agua

water temperature switch, interruptor de temperatura de agua

waterflow switch, interruptor de flujo de agua

watertight, impermeable al agua

watt, vatio

wattage, vataje

watt-hour meter, medidor vatios/hora

watt-hours, vatios-hora

wattmeter, vatímetro

watts per lineal foot, vatios por pie lineal

wave, onda

waveform, forma de onda

weather resistant, resistente a la intemperie

weatherproof, a prueba de intemperie

weatherproof covers, tapas a prueba de intemperie

weight, peso

wet-standpipe system, sistema de toma de agua húmedo

wide/width, ancho/anchura

winding, embobinado

wire, alambre

wire cutters, cortaalambres

wire gauge size, calibre del cable

wire nuts, tuercas para alambre

wire strength, resistencia del alambre

wire support, soporte de alambre

wireframe, bastidor

wireless communication, comunicación inalámbrica

wireway, tendido de cables

wiring method, método de cableado

wiring system, sistema de cableado

with no current limitation, sin limitación de corriente

wood screw, tornillo para madera

wooden plug, clavija de madera

working space, lugar de trabajo

wound, embobinado

Y

yoke, yugo

Z

zinc, zinc

SPANISH-ENGLISH
ESPAÑOL-INGLÉS

A

a prueba de explosión, explosion-proof

a prueba de golpes, taper proof

a prueba de ignición-polvo, dust-ignition proof

a prueba de intemperie, weatherproof

a prueba de polvo, dustproof

a prueba de salpicaduras, drip proof

abertura de una caja, box opening

abierto, open

abrasión, abrasion

abrazadera para manguera flexible, flexible hose clamp

abrazaderas de haz, beam clamps

acabado metálico exterior e interior, exterior and interior metallic finish

accesible, accessible

acceso de lógica, logic gate

accesorio ER, ER fitting

accesorio incandescente empotrado, recessed incandescent fixture

accesorios para iluminación de emergencia, emergency lighting fixtures

acero estructural, structural steel

acero galvanizado, galvanized steel

acometida de servicio, service head

acoplamiento, coupling

acoplamiento de PVC, PVC coupling

acoplamiento metálico flexible "Pig Tail", Pig tail

acople sin rosca, threadless coupling

acoples roscados, threaded couplings

acortamiento, shortening

activación de comunicaciones de voz de emergencia, activation of emergency voice communications

activación del dispositivo de inicio, activation of the initiating device

activar y desactivar la corriente del circuito, start and stop circuit current

actualizar, update

acumulador, storage battery

Administración, administration

advertencias eléctricas, electrical warnings

afiladora angular, angle grinder

afiladora recta, die grinder

aguja, needle

aguja indicadora, pointer

agujero de inspección, handhole

aislado, insulated

aislador, insulator

aislador de calor, heat insulator

aislador térmico, heat sink

aisladores termoencogibles, heat-shrink insulators

aislamiento, insulation

aislamiento del conductor neutro, neutral conductor insulation

aislamiento del conector puesto a tierra, grounded conductor insulation

aislamiento refractario altamente comprimido, highly compressed refractory insulation

aislamiento térmico, thermal insulation

aislante de vibración, vibration isolator

ajustar, tight

ajuste por reducción, deduction adjustment

al ras de la superficie terminada de la pared, flush with the finished face of the wall

alambre, wire

alambre guía, fish tape

alambres cruzados, crossed wires

alambres de soporte independientes, independent support wires

alargamiento, elongation

alarma de termómetro, thermometer alarm

alarma sonora, audible alarm

aleación de cobre, copper alloy

alimentador, feeder

alimentador aéreo, overhead feeder

alta tensión/alto voltaje, high voltage

alterna, alternating

alto, high

altura, height

aluminio, aluminum

aluminio revestido con cobre, copper-clad aluminum

ampacidad del alimentador, feeder ampacity

amperaje, ampacity

amperaje de los cables, cable ampacity

amperaje de transporte de corriente, current carrying ampacity

amperaje nominal, rated ampacity

amperímetro, ammeter

amperímetro de pinza, clamp-on ammeter

amperio, ampere

amperio-hora, amp-hour

amperios del rotor enclavado, locked rotor amperes

amps, amps

análisis de esfuerzos, stress analysis

analizador de tensión, voltage tester

ancho/anchura, wide/width

anclajes de tornillos de expansión, expansion screw anchors

anillo de masa, ground ring

anillo de yeso, plaster ring

antena, antenna

antena parabólica, satellite dish

anunciador, annunciator

apriete, grip

aproximación, proximity

apuntalamiento, shoring

apuntalamiento de zanjas, shoring of trenches

arandela aislante, grommet

arandela de goma, rubber washer

arco, arcing

arco eléctrico, electric arc

área de la sección transversal, cross-sectional area

área transversal, cross sectional area

área transversal de ocupación total, total cross-sectional fill area

áreas de almacenamiento, storage areas

argón, argon

armar un circuito, build a circuit

armónico, harmonic

arrancador de motor, motor starter

arrancador magnético, magnetic starter

arrastre, drag

artefactos de luces de salida, exit lighting fixtures

asa, handle

atenuar, dim

aumentar, increase

automatización dura, hard automation

autotransformador de vataje constante, constant-wattage - auto-transformer

AWG, AWG

B

baja energía en las baterías, low battery power

bajada general, overhead drop

bajadas de suministro de comunicación, communication supply drops

bajo voltaje, under voltage

bakelita reforzada con fibra, fiber-reinforced bakelite

balasta, ballast

balasto de accesorio, fixture ballast

bandeja portacables, cable tray

barra conductora, busbar

barra de refuerzo de acero, steel reinforcing bar

base abisagrada, hinged base

base Edison estándar, standard Edison-base

bastidor, wireframe

batería, battery

batería de larga vida, long-life battery

batería de plomo-ácido, lead acid battery

bidimensional, two dimensions

bifurcación con cables múltiples, multi-wire branch

biseladora, beveler

blindado, shielded

blindaje de cables, cable armor

bloque de concreto, concrete block

bloquear, lock

bloqueo/etiquetado de energía, lockout/tagout

bloqueos de canal, channel locks

bloques de mampostería, masonry block

bobina, coil

bobina de alambre, coils of wire

bobina primaria, primary coil

boceto eléctrico, electrical rough-in

bomba, pump

bomba contra incendio, fire pump

bomba de circulación de agua caliente, hot water circulating pump

bomba de irrigación, irrigation pump

bombilla eléctrica, light bulb

borde, edge

borde burdo, rough edge

bóveda, vault

brazo robot, robotic arm

brecha/hueco, gap

broca, drill bit

bruñir, burnish

bucle-arranque, loop-start

buje aislado, insulated bushing

C

caballo de fuerza, horsepower

caballos de fuerza nominales, horsepower rating

cabeza terminal de cable, pothead

cable a tierra, ground wire

cable adicional, extra cable

cable blindado, armored cable

cable coaxial, coaxial cable

cable con tensión a tierra, hot wire to the ground

cable con vaina de cobre, copper-sheathed cable

cable con vaina no metálica, nonmetallic sheathed cable

cable de conductores axiales retorcidos, twinax

cable de conexión, jumper cable

cable de empalme, bond wire

cable de prueba, test lead

cable de tipo FPLP, FPLP type cable

cable de tipo NM, type NM cable

cable de UF, UF cable

cable de un par blindado, single pair shielded

cable flexible, flexible cable

cable mineral aislado (MI), mineral insulated cable (MI)

cable multiconductor, multiconductor cable

cable neutro, neutral wire

cable no metálico, nonmetallic cable

cable PFL, FPL cable

cable plano de múltiples conductores, multiconductor ribbon cable

cable portador, messenger wire

cable THHN, THHN wire

cable tipo MC, type MC cable

cable tipo MI, type MI cable

cableado de campo ignífugo, nonincendive field wiring

cableado eléctrico, electrical wiring

cableado general, general wiring

cableado subterráneo, underground wiring

cables con cubierta no metálica, non-metallic sheathed cable

cables de tipo a tierra, grounding extension cords

cables enterrados, buried cables

cabriada, joists

cadena de seguridad, safety chain

caída, drop

caída de servicio, service drop

caída de tensión combinada, combined voltage drop

caída de voltaje, voltage drop

caja de acceso manual, manual pull box

caja de clavijas de pared, wall plug box

caja de conexiones, junction box

caja de conexiones de bronce, brass junction box

caja de dispositivos, device box

caja de fusibles, fuse box

caja de salida, outlet box

caja de salida eléctrica, electrical outlet box

caja en campana, bell box

caja NEMA, NEMA enclosure

caja octagonal, octagon outlet box

caja plana, square box

caja sin empalme o derivación, box without a splice or tap

cajas en obras ocultas, boxes in concealed work

calentador, heater

calentador de sobrecarga, overload heater

calibrador de roscas, thread gauge

calibre, gauge

calibre del cable, wire gauge size

calor, heat

calor del arco, heat of the arc

cámara, chamber

cámara con circulación de aceite, oil-filled chamber

cámara de distribución, plenum

cámaras con circulación de gas, gas-filled chambers

campo magnético, magnetic field

canal de descarga de arco, arc chute

canaleta, raceway

canaleta de metal, metal raceway

canaleta de tránsito, busway

canaleta embutida, flush raceway

canaleta ferrosas, ferrous raceway

canalón, gutter

cañería eléctrica metálica, electrical metallic tubing

capacidad de corte, interrupting rating

capacidad de irrupción promedio, average inrush capacity

capacidad de supervivencia, survivability

capacidad de transporte de corriente, current-carrying capacity

capacidad interrumpida disponible, available interrupted capacity

capacitancia, capacitance

capacitivo, capacitive

capacitor para funcionamiento de motores, motor run capacitor

carcasa, shell

carga, charge

carga, load

carga alta, high-load

carga calculada, computed load

carga continua, continuous load

carga de iluminación, lighting load

carga del alimentador, feeder load

carga eléctrica, electrical load

carga estática de electricidad, static charge of electricity

carga máxima, maximum charge

carga momentánea, momentary load

carga resistiva, resistive load

carga servida, load served

cargador de baterías, battery charger

carrera/desplazamiento, stroke

carrete, reel

cartón prensado, pressed board

cartucho, cartridge

casco del tornillo, screw shell

cautín, soldering iron

centro de control de motores, motor control center

cepillo para conductos, conduit brush

cesta de arrastre, pulling basket

ciclo de lectura y escritura, read and write cycle

cielorraso, ceiling

cierre de seguridad, safety latch

cierre por torsión, twist-lock

cincel, cold chisel

cinta de electricista, electrician's tape

cinta de medir, measuring tape

cinta de recambio, replacement link

circuito, circuit

circuito abierto, open circuit

circuito aislado, isolated circuit

circuito combinatorio, combination circuit

circuito de alimentación, feeder circuit

circuito de alimentación y derivación, feeder and branch circuit

circuito de CA, AC circuit

circuito de comunicaciones, communications circuit

circuito de control remoto, remote-control circuit

circuito de derivación, branch circuit

circuito de línea de señalización, signaling line circuit (SLC)

circuito de múltiples derivaciones, multi-branch circuit

circuito de señales, signal circuit

circuito derivado de un motor, motor branch circuit

circuito derivado residencial, residential branch circuit

circuito eléctrico, electrical circuit

circuito en paralelo, parallel circuit

circuito en serie, series circuit

circuito excitador, drive circuit

circuito experimental, breadboard

circuito integrado, integrated circuit

circuito mixto/combinado, series-parallel circuit

circuito multicableado, multiwire circuit

circuito sin potencia limitada, non-power limited circuit

circuito temporal, temporary circuit

circunferencia, circumference

circunferencia del conducto, circumference of the conduit

clavija, plug

clavija de conexión, attachment plug

clavija de madera, wooden plug

cloruro polivinílico, polyvinyl chloride

coaxial, coax

cobre, copper

cobre revestido en níquel, nickel-clad copper

código alfanumérico, alpha-numeric code

código de falla del diagnóstico, diagnostic trouble code

código eléctrico nacional, national electrical code

codo concéntrico, concentric bend

codo de barrido, sweep bend

codo de campo, field bend

codo en caballete, saddle bend

codo horizontal, horizontal elbow

codo radial, radius bend

codo vertical, vertical elbow

codos, bends

colocación de flejes, strapping

combinación de clavija y tomacorriente, plug and receptacle combination

compartido, party-line

compartimento con humo, smoke compartment

compensar, compensate

componente, component

componente activo, active component

componente pasivo, passive component

compresor de aire acondicionado, air conditioning compressor

comprobador de línea óptico, optical line tester

compuesto de arrastre, pulling compound

compuesto desoxidado, de-oxide compound

comunicación inalámbrica, wireless communication

comunicaciones en red, network communications

con protección térmica, thermally protected

concentrador Myers, Myers hub

concreto vaciado, poured concrete

condensador, capacitor

condensador, condenser

condición cerrada, lock condition

condición de falla a tierra, ground fault condition

condiciones destructivas, destructive conditions

condiciones normales, standard conditions

conductividad, conductivity

conducto de PVC, PVC conduit

conducto eléctrico de metal rígido, electrical rigid metal conduit

conducto expansible, expandable conduit

conducto galvanizado, galvanized conduit

conducto IMC y EMT rígido, rigid IMC and EMT conduit

conducto metálico, metal conduit

conducto metálico flexible, flexible metal conduit

conducto metálico rígido, rigid metal conduit

conducto no caldeado, unheated conduit

conducto no metálico, nonmetallic conduit

conductor, conductor

conductor aislado, insulated conductor

conductor con barril aislado con goma, rubber-insulated barrel connector

conductor de aluminio, aluminum conductor

conductor de cobre, copper conductor

conductor de cobre desnudo, bare copper conductor

conductor de fase B, B-phase conductor

conductor de fase caliente, hot phase conductor

conductor de la entrada de servicio, service entrance conductor

conductor de puesta a tierra, grounding conductor

conductor de salida y retorno, outgoing and return conductor

conductor de servicio conectado a tierra, grounded service conductor

conductor del circuito, circuit conductor

conductor del circuito conectado a tierra, grounded circuit conductor

conductor del electrodo conectado a tierra, grounded electrode conductor

conductor del electrodo de puesta a tierra, grounding electrode conductor

conductor del ramal de acometida, service-drop conductor

conductor no conectado a tierra, ungrounded conductor

conductor tipo THHN, type THHN conductor

conductor trenzado, stranded conductor

conductores de derivación, branch circuit conductors

conductores de suministro, supply conductors

conductores vivo y neutro, hot and neutral conductors

conector, connector

conector de clavija, plug connector

conector de dos patas, two post connector

conector de perno dividido, split-bolt connector

conector de tipo de compresión irreversible, irreversible compression-type connector

conector Romex, Romex connector

conectores de empalme a tope, butt splice connectors

conexión a una canaleta, connection to a raceway

conexión alámbrica, tie wire

conexión de conductos compactos, compact conductor connection

conexión de conductos de carga, load conductor connection

conexión de conductos desnudos, bare conductor connection

conexión de empalme giratoria, swivel joint connection

conexión flexible, flexible connection

conexión floja, loose connection

conexión soldada, soldered connection

configuración, setup

configuración variable, variable setting

conjunto de clavijas, plug set

conjunto de generador, generator set

conjunto de múltiples salidas, multi-outlet assembly

conjunto integrado, packaged assembly

conmutador rotativo, rotary switch

constante de tiempo de circuito RC, RC time constant

construcción de rotor, rotor construction

contacto de tierra en U, U-ground contact

contacto eléctrico, electrical contact

contacto fijo, stationary contact

contacto móvil, movable contact

contactos de hoja, blade contacts

continuidad eléctrica, electrical continuity

continuo, continuous

contorno, outline

contracción térmica, thermal contraction

contratuerca, locknut

control de flujo, flow control

control estadístico del proceso, statistical process control

control numérico por computadora, computer numerical control

control regulador de motor, motor control

controlador de motor, motor controller

controlador de motor tipo 3R NEMA, NEMA Type 3R motor controller

controlador del sistema de aspersión, sprinkler system controller

controlador lógico programable, PLC

controlador lógico programable, programmable logic controller

conversor de fase, phase converter

convertidor de voltaje/tensión, voltage converter

correa de instalación, mounting strap

correa de un orificio, one-hole strap

corriente, current

corriente a plena carga, full-load current

corriente alterna (CA), alternate current (AC)

corriente alterna inducida, alternating induced current

corriente alterna rectificada, rectified alternating current

corriente de carga, load current

corriente de cortocircuito, short circuit current

corriente de falla, fault current

corriente de funcionamiento, operating current

corriente de marcha, running current

corriente de sobrecarga, overload current

corriente de soldadura de alta tensión, high voltage welding current

corriente del rotor enclavado, locked-rotor current

corriente directa (CD), direct current (DC)

corriente eléctrica, electrical current

corriente inductiva, inductive current

corriente nominal, rated current

corriente parásita, stray current

corrientes de tierra, ground currents

corrosión, corrosion

cortaalambres, wire cutters

cortacaños, pipe cutter

cortacircuito magnético, magnetic cutout

cortacircuitos, circuit breakers

cortafuego, fire stopper

cortar, interrupt

cortocircuito, short circuit

cristal, crystal

cuarzo, quartz

cubierta, cover

cubierta no combustible, non-combustible cover

cubierta no metálica, nonmetallic cover

cubierta para cables, cable sheath

cubierta protectora, cover plate

cubierta protectora, housing

cubiertas metálicas puestas a tierra, grounded metallic covers

cubo roscado, threaded hub

cubo sellador, sealing hub

cuello de cisne, gooseneck

cuerda, pitch

cuerpo conductor, conducting body

CH

chicote, stub-up

chispas, sparking

choque eléctrico, electric shock

choque térmico, thermal shock

cumplimiento, enforcement

cúpula, canopy

D

de acción automática, self-acting

decibeles, decibels

del polo negativo a tierra, negative pole to ground

del polo negativo al positivo, negative pole to positive pole

del polo positivo a tierra, positive pole to ground

del polo positivo al negativo, positive to negative pole

depresión, depression

derivación, branch/shunt

desbarbar, de-burr

descarga de alta intensidad, high intensity discharge

descarga de ión de arco de helio, heliarc ion discharge

descenso a velocidad constante, constant rate descent

desmagnetizar, demagnetize

desmontar, disassemble

destornillador, screwdriver

desviación, offset

detector de humo, smoke detector

detector de llama, flame detector

detector de temperatura, temperature detector

detector de velocidad de aumento, rate of rise detector

diagnosticar, diagnose

diagnóstico, diagnosis

diagrama, diagram

diagrama ascendente, riser diagram

diagrama de cableado entre puntos, point to point wiring diagram

diagrama de línea única, single line diagram

diámetro, diameter

diámetro del orificio, diameter of hole

diámetro externo, outside diameter

diámetro interior, inside diameter

diferencia, difference

dimensión, dimension

diodo, diode

diodo emisor de luz (LED), light emitting diode (LED)

directamente proporcional, directly proportional

disipador de sobretensión, surge arrester

disminución de potencia, derating

disparar, blow

disparo fijo, fixed trip

dispositivo, device

dispositivo automático activado por la luz, automatic light-actuated device

dispositivo de alarma contra incendio, fire alarm device

dispositivo de arranque, starter

dispositivo de auto-monitoreo, self-monitoring device

dispositivo de iniciación, initiation device

dispositivo de medición, metering device

dispositivo de notificación visible, visible notification appliance

dispositivo de sobrecorriente, overcurrent device

dispositivo eléctrico, electrical device

dispositivo electromagnético, electromagnetic device

dispositivo fluorescente colgante, pendant-type fluorescent fixture

dispositivo incandescente, incandescent fixture

dispositivo lleno de aceite, oil-filled device

dispositivo para leer los amperios, ampere reader ammeter

dispositivo que funciona a motor, motor operated appliance

dispositivo sellado amorfo, amorphous sealed device

dispositivo sensible a la polaridad, polarity-sensitive device

dispositivos audibles, audible appliances

dispositivos de conmutación o control, switching or control devices

distancia, distance

distancia de centro a centro, center to center distance

distancia inicial, input distance

distancia máxima, maximum distance

distancia mínima, minimum distance

distancia resultante, output distance

distribución, distribution

distribución de yodo halógeno, halogen iode distribution

distribución industrial pesada, heavy industrial distribution

disyuntor, breaker

disyuntor de circuito en caja moldeada, molded-case circuit breaker

disyuntor de cuchilla, knife disconnect

disyuntor de falla de arco, arc fault breaker

disyuntor de tiempo inverso, inverse time circuit breaker

dobladora hidráulica, hydraulic bender

dobladora manual, hand bender

doblar/curvar, bend

doble servo, duo-servo

dureza, hardness

E

ecuación de la ley de ohm, ohm's law equation

eje motor, drive axle

electricidad, electricity

electricista, electrician

electrochoque, electrical shock

electrodo de puesta a tierra, grounding electrode

electrodo empotrado en concreto, concrete-encased electrode

electroimán, electromagnet

electrólito, electrolyte

electromagnetismo, electromagnetism

electromecánico, electromechanical

electrostático, electrostatic

elevación original, original elevation

elevar o reducir la tensión, step-up or step-down of voltage

embobinado, winding

embobinado, wound

emisión de humos tóxicos, release of toxic fumes

emisor de luz, emitter of light

empalmado, bonded

empalme, bonding

empalme de expansión, expansion joint

empotrar, encase

EMT (tuberías metálicas eléctricas), EMT

en bucle, looped

en paralelo, in parallel

en serie, in series

encendido electrónico, electronic ignition

energía alternativa, alternative energy

energía calórica, heat energy

energía de emergencia, emergency power

energía eléctrica, electrical power

energía renovable, renewable energy

enrollado en espiral, pig tailed

entalladura, notch

entibamiento, banking

entrada, input

entrada al motor, input to the motor

entretecho, attic

envoltura del cable, cable jacket

equipo, equipment

equipo de conserjería, janitorial equipment

equipo de control de seguridad, safety-control equipment

equipo de descongelamiento, de-icing equipment

equipo de distribución, switchgear

equipo de prueba, test equipment

equipo eléctrico, electrical equipment

equipo receptor, receiving equipment

escala de resistividad, resistivity scale

escalera, ladder

escalera removible, removable ladder

escariador cónico, tapered reamer

escariar, ream

esfuerzo, stress

espaciador, spacer

espacio aéreo, air space

espacio eléctrico, electrical space

espacio hueco, hollow space

especificaciones, specifications

especificaciones del motor, motor specifications

especificar, specify

esquema, schematic

estabilizar, stabilize

estación de alimentación, supply station

estación de arranque/parada, start/stop station

estación de supervisión remota, remote supervising station

estado sólido, solid-state

estereoscópico, stereo optic

estrella y caja, star and box

estructura adyacente, adjacent structure

etiqueta/identificación, tag

etiquetas de precaución, caution tags

exotérmico (soldadura de cadmio), exothermic (cadwelding)

exponencial, exponential

expuestos al aceite y la gasolina, exposed to oil and gasoline

extensión flexible, flexible cord

exterior, outdoors

extremo, end

extremo de corte, cut end

F

fácil de localizar, able to locate

factor de ajuste, adjustment factor

factor de carga, load factor

factor de demanda, demand factor

factor de potencia, power factor

factor de resistencia, resistance factor

falla, failure

falla, fault

falla de tierra, ground fault

falla eléctrica, electrical failure

faradios, farads

fasímetro, phase meter

fertilizante, fertilizer

fibra, fiber

fibra de vidrio, fiberglass

fibra óptica, fiber optics

fibra óptica, optical fiber

fijadores atravesados ("molly"), molly-or toggle-type fasteners

filamentos por pulgada, filament

fluctuación de voltaje/tensión, fluctuating voltage

fluctuaciones de corriente, current to fluctuate

flujo continuo, continuous flow

flujómetro, flowmeter

fluorescente, fluorescent

forma de onda, waveform

fotovoltaico, photo-voltaic

frecuencia, frequency

frecuencia resonante, resonant frequency

frecuencia variable, varying frequency

fricción, friction

fuente alterna de corriente, alternate source of power

fuente alterna de energía, energy alternate sources

fuente de energía, power source

fuente de energía alterna, alternate energy

fuente de energía de emergencia, emergency power source

fuente de energía eléctrica, electrical power supply

fuerza aplicada, input force

fuerza electromotriz, electromotive force

fuerza magnética, force of magnetism

fuerza magnética, magnetic strength

fundamentos de lógica digital, digital logic fundamentals

fundirse un fusible, blow a fuse

fusible, fuse

fusible de cortocircuito/falla a tierra, short-circuit/ground fault fuse

fusible de dos elementos, dual element fuse

fusible de paso, step fuse

fusible de retardo, time-delay fuse

fusible restaurable, resettable fuse

fusible tipo S, Type S fuse

G

gabinete, cabinet

galón, gallon

gancho aéreo, sky hook

gancho para levantar, lifting hook

gas hidrógeno, hydrogen gas

gel lubricante, lubricating gel

generador, generator

generador de puesta a tierra portátil, grounding portable generator

generador de reserva de emergencia, emergency standby generator

generador montado sobre vehículos, vehicle-mounted generator

grado de curvatura, degree of bend

grados, degrees

grapa, staple

grasa de cañerías, pipe grease

gravitacional, gravity operated

grosor, thickness

guía de deslizamiento, slideway

H

halogenuro metálico, metal halide

haz de electrones, electron beam

helio, helium

herramienta especial, special tool

herramienta hidráulica, hydraulic tool

herramienta magnética, magnetic tool

herramienta manual, hand tool

herramienta neumática, pneumatic tool

herramienta para corte de metal, metal cutting tool

hertzios (Hz), hertz (Hz)

HID, HID

hilo conductor del motor, motor lead wire

hilo exterior del interruptor, switch leg

hisopo para conductos, conduit swab

humedad, moisture

I

identificado, identified

iluminación de carnaval, carnival ride lighting

iluminación de realce, festoon lighting

iluminación paisajística, landscape lighting

iluminación subacuática, underwater lighting

iluminar, illuminate

imán, magnet

imantación, magnetization

impedancia, impedance

impermeable a la lluvia, raintight

impermeable a los líquidos, liquid tight

impermeable al agua, watertight

incendio eléctrico, electrical fire

indicación EOL, designation EOL

indicador, indicator

indicador de luz de detención, stall light indicator

indicador de reloj, dial indicator

indicador principal de energía, main power indicator

inducción, induction

inductancia, inductance

inductivo, inductive

inductor, inductor

inflamable, flammable

información sobre detección de fallas, troubleshooting information

ingeniero eléctrico, electrical engineer

instalación, installation

instalación bajo tierra, burial

instalación con múltiples servicios, installation with multiple services

instalación de bandeja portacables, cable tray installation

instrumentos virtuales, virtual instrumentation

interbloqueo eléctrico, electrical interlock

intercambiable, interchangeable

intercambiados, crossed

interfaz de circuitos, circuit interface

interferencia electromagnética, electromagnetic interference

interrupción a tierra, open ground

interruptor, switch

interruptor a nivel del agua, water level switch

interruptor accionado por corriente de pérdida a tierra, ground fault circuit interrupter (GFCI)

interruptor aislador, isolating switch

interruptor bidireccional unipolar, single-pole double-throw (SPDT) switch

interruptor de botón pulsador doble, dual pushbutton switch

interruptor de circuito de falla ARC, ARC fault circuit interrupter

interruptor de cuchilla unidireccional, single-throw knife switch

interruptor de flotador, float switch

interruptor de flujo de agua, waterflow switch

interruptor de inserción de fallas, fault insertion switch

interruptor de presión, pressure switch

interruptor de servicio, service switch

interruptor de temperatura de agua, water temperature switch

interruptor de transferencia, transfer switch

interruptor de transferencia de emergencia, emergency transfer switch

interruptor eléctrico de pared, electrical wall switch

interruptor limitador, limit switch

interruptor neutral de seguridad, neutral safety switch

interruptores manuales exteriores, exterior manual switches

intervalo, interval

inversamente proporcional, inversely proportional

isométrico, isometric

J

jaula de ardilla, squirrel cage

junta sellada con goma de silicón, silicone rubber sealed joint

K

kilovatio-hora, kilowatt-hour

kilovatios (kW), kilowatts (kW)

kilovoltios-amperios (kVA), kilovolts-amperes (kVA)

L

lámpara, lamp

lámpara fluorescente, fluorescent lamp

lámpara halógena de tungsteno, tungsten halogen lamp

lámpara tubular de halógeno, halogen tubular lamp

láser modulado, modulated laser

lazos de goteo, drip loops

lectura precisa, accurate reading

legibilidad, readability

lengüeta, tab

libras, pounds

libre de polvo, dust-free

libre de tierra, ground-free

lijar/arena, sand

limitación de potencia, power-limited

línea, line

línea de campo, field line

linealidad, linearity

lista de verificación, check list

llave de cadenas, chain wrench

llave hexagonal, hex key

lógica digital, digital logic

lógica en escalera, ladder logic

lubricante en atomizador, spray lubricant

luces de emergencia, emergency lighting

luces de neón, neon lighting

lugar de trabajo, working space

lugares corrosivos, corrosive locations

lumen, lumen

luz estroboscópica roja titilante, flashing red strobe light

M

magnetita, loadstone

magnetita, magnetite

mal funcionamiento del medidor, meter malfunction

malla, net

mampostería/albañilería, masonry

manguito de expansión, lag shield

manómetro, manometer

máquina de análisis de humo, smog testing machine

maquinaria impulsada, driven machinery

marco del motor, motor frame

marco metálico de la estructura, metal frame of the structure

martillo, hammer

martillo de agujas, needle scaler

martillo de bola, peen hammer

martillo de plástico, plastic hammer

martillo remachador, riveting hammer

material aislante, insulating material

material combustible, combustible material

material inflamable, flammable material

material retardador de incendios, fire-retardant material

material Unistrut, Unistrut material

matriz, matrix

mecanismo de transmisión, transmission mechanism

medición lineal, linear measurement

medición transversal, cross-sectional measurement

medida de centro a centro, center to center measurement

medida de la eficiencia del motor, measure of motor efficiency

medidor de entrada, meter-in

medidor de salida, meter-out

medidor vatios/hora, watt-hour meter

medidor/metro, meter

medios de desconexión, disconnecting means

megaóhmetro, megger

megaóhmetro, megohmmeter

megavatios (MW), megawatts (MW)

ménsula/soporte, bracket

metal corrugado, corrugated metal

metal oxidado, corroded metal

método de cableado, wiring method

método de cableado fijo, fixed wiring method

metro cuadrado, square meter

micrómetro, micrometer

microplatina, microslide

miembro estructural, structural member

miliamperímetro, milliameter

miliamperios (ma), milliamps

milipulgada circular, circular mil

mínimo, minimum

modo de conversación, talk mode

módulo terminal del motor, motor terminal module

molde de "cadweld", cadweld mold

monofásico, one-phase

monofásico, single phase

montar, mount

mordaza de Kellum, Kellum's Grip

mosfet de potencia, power mosfet

motor de corriente directa, DC motor

motor de inducción, induction motor

motor de tipo universal, universal type motor

motor eléctrico, electric motor

motor en jaula de ardilla, squirrel cage motor

motor en jaula de ardilla de CA, AC squirrel cage motor

motor estacionario, stationary motor

motor neumático, air motor

motor polifásico CA, AC polyphase motor

movimiento de traslación, traslational motion

movimiento hacia delante, forward motion

movimiento rectilíneo, linear motion

multímetro (VOM), multimeter (VOM)

multímetro digital, digital multimeter

N

negativo, negative

neón, neon

nicho, niche

ningún equipo puesto a tierra, no equipment ground

niple, nipple

níquel, nickel

nivel de sonido ambiental, ambient sound level

norma, standard

Normas de Calibración Norteamericanas, American Wire Gauge Standards

normatividad de seguridad, safety regulations

núcleo, core

núcleo de hierro, iron core

número de libras por pie de cable, number of pounds per foot of wire

número de roscas por pulgada, number of threads per inch

O

oblicuo, skewed

obligatorio, mandatory

obstrucción, obstruction

ocupación de una caja, box fill

ohm, ohm

ohmímetro, ohmmeter

onda, wave

onda electromagnética, electromagnetic wave

onda sinusoidal, sine wave

operación, operation

operado por solenoide, solenoid-operated

operativo, operational

opto acoplador, opto-coupler

orejeta reglamentaria, listed lug

orientador de antena, disk-aimer

orificio, hole

oscilador, oscillator

osciloscopio, oscilloscope

osciloscopio de doble traza, dual trace oscilloscope

oxidación, rusting

oxidado, corroded

P

palanca, toggle

panel, board

panel de control, panelboard

par de torsión de arranque, starting torque

par torcido, twisted pair

paralaje, parallax

paralelo, parallel

pararrayos, lightning arrester

paredes combustibles y no combustibles, combustible and non-combustible walls

parte con voltaje/tensión, live part

parte conductora, conductive part

parte oxidada, rusted part

partículas de combustión en el aire, combustion particles in the air

partículas de protones, proton particles

pasador cónico, taper pin

pasamuro, bushing

pasta antioxidante, anti-oxidant paste

patrón, pattern

pegamento, glue

peldaño aislado, insulated step

peldaños antideslizantes, non-skid feet

pendiente, hanging

penetración eléctrica, electrical penetration

pérdida del transformador, transformer loss

período, period

perno dividido, split bolt

pernos Rawl, Rawl-drives

peso, weight

pies, feet

pintura resistente a la luz ultravioleta, ultra-violet resistant paint

pinza de cables, cable clamp

pinzas de punta de aguja, needle nose pliers

piso de tipo celular, cellular floor

pivote, pivot

placa "citrus", citrus plate

placa de aplicaciones del microprocesador, microprocessor applications board

placa de barrera, barrier plate

placa de empalme, splice plate

placa de hierro o acero, plate of iron or steel

placa de montaje del motor, motor mounting plate

placa de soporte, base plate

placa giratoria, swivel plate

placas de empalme horizontal, horizontal splice plates

placas metálicas puestas a tierra, grounded metallic plates

plano de techo reflejado, reflected ceiling plan

plano de ubicación, site plan

plano mecánico, mechanical plan

plástico térmico, thermoplastic

plataforma, platform

plataforma/cubierta, deck

platina, slide

plomada, plumb-bob

plomo, lead

polaridad, polarity

polaridad inversa, reverse polarity

polea, pulley

polímeros extruidos, extruded polymers

polipasto neumático, pneumatic hoist

poliuretano, polyurethane

polo consecutivo, consequent pole

polo salientes, salient-pole

porcentaje, percent

porcentaje de deslizamiento, percent slip

porta-accesorios, fixture stud

porta-carretes, reel jack

portafusibles, fuse holder

porta-herramientas, tool holder

portalámpara, lampholder

portátil, portable

posicionamiento incremental, incremental positioning

positivo, positive

poste metálico, metal pole

potencia, power

potenciómetro, potentiometer

pozo, pit

precableado, pre-wired

preciso, accurate

preevaluación, pretest

prensa, press

prensa hidráulica, hydraulic press

prensa mecánica, mechanical press

prensa neumática, pneumatic press

presilla de puesta a tierra, grounding clip

prevención de la diseminación de humo, smoke spread prevention

principio de transformador de corriente, current-transformer principle

probador de continuidad de audio, audio continuity tester

procesador digital de señales, digital signal processor (DSPS)

profundidad, depth

protección anticorrosiva, corrosion protection

protección contra cortocircuitos, short circuit protection

protección contra falla a tierra, ground fault protection

protección contra sobrecarga, overload protection

protección contra sobrecorriente, overcurrent protection

protector térmico, thermal protector

prueba de reaceptación, reacceptance test

prueba preliminar, preliminary check

psi, psi

puente de conexión a tierra de servicio, service bonding jumper

puente de conexión flexible, flexible bonding jumper

puente principal, bonding jumper

puente principal, main bonding jumper

puerto en serie, serial port

puesta a tierra Ufer, Ufer ground

puesta en funcionamiento, start

puesto de tracción, pull station

pulgada cúbica, cubic inch

pulgadas, inches

pulgadas cuadradas, square inches

pulsador del motor, motor pushbutton

pulsante, pulsating

punto de acceso, pull point

punto de conexión, point of attachment

punto de conexión de un ramal de acometida, attachment point of a service drop

punto de fusión, melting point

punto de ignición, ignition point

punto de prueba, test point

Q

químicos alcalinos, alkali chemicals

R

radiación, radiation

radiación infrarroja, infrared radiation

radio de los cables, radius of the wires

radiodifusión, radio broadcasting

rama de seguridad de vida, life-safety branch

rango de entrada automático, auto-ranging

ranura superficial en concreto, shallow chase in concrete

rasgar, scar

reacción endotérmica, endothermic reaction

reacción exotérmica, exothermic reaction

reacción química, chemical reaction

reactor, reactor

realce, festoon

rebajador, step-down

receptáculo, receptacle

recinto, enclosure

recinto de los equipos de transferencia, transfer equipment enclosure

recinto para tomacorriente, receptacle enclosure

recorrido, span

recorrido de un conducto, conduit run

rectificador, rectifier

red, network

reductor, dimming

reductor, kick

reemplazar, replace

referencia a símbolos, symbol legend

régimen, rating

régimen de caballos de fuerza, rated horsepower

régimen del conductor común, rating of the bussing

régimen NEMA, NEMA rating

régimen total de amperios de las unidades, total ampere ratings of the units

regla de la toma del alimentador, feeder tap rule

regulación de frecuencia, setting frequency

regulador, buffer

regulador, regulator

regulador de voltaje, voltage regulator

regular la velocidad, regulate speed

reinstalar, reinstall

rejillas de techo, ceiling grids

relé de retardo de tiempo, time-delay relay

relé de sobrecarga, overload relay

relé/relevador, electric relay

relé/relevador, relay

remolcador, tugger

removible, removable

reóstato, rheostat

reparaciones, servicing

reposición, reset

reprogramar, reprogram

requisitos de conexión a tierra, grounding requirements

requisitos de espaciado, requirements of the spacing

requisitos del ADA (Americans with Disabilities Act) para instalaciones eléctricas, ADA handicapped electrical installation requirements

resistencia, resistance

resistencia de aislamiento, insulation resistance

resistencia del alambre, wire strength

resistente a la intemperie, weather resistant

resistente al calor, heat resistant

resistor, resistor

resistor variable (RV), variable resistor (V.R.)

retardador de llama, flame-retardant

retornos a tierra, earth returns

retranqueo, set back

retraso, time-delay

revestimiento, envelope

revestimiento no conductor, nonconductive coatings

robot trazador, tracer robot

rodear, encircle

rodillo, roller

rodillo alimentador, feed roller

rosca, thread

roscado, threaded

roscas por pulgada, threads per inch

rótulo, nameplate

rótulo del motor, motor nameplate

rueda de un neumático, tire rim

ruido eléctrico, electrical noise

ruta, path

S

sala eléctrica, electrical room

salas de acumuladores, storage battery rooms

salida, output

salida del motor, output of the motor

sección transversal combinada de los cables, combined cross-sectional area of the wires

secuencia, string

secuencia de operaciones de control, sequence of control operations

secuencial, sequencing

seguridad, safety

seguridad y salud en el trabajo, occupational safety and health

sello a prueba de lluvia, raintight seal

semiautomático, semi-automatic

semiconductor, semi-conductor

sensor de calor, heat sensor

señal de bocina del estrobo, horn strobe

señal ultravioleta, ultra violet signature

señales digitales, digital signals

señales eléctricas, electrical signals

señalización, signaling

servo sincronizado, synchro- servo

servo-mecanismo, servomechanism

sierra de calar, keyhole saw

sierra para metales, hacksaw

símbolo esquemático, schematic symbol

sin aislamiento, bare

sin limitación de corriente, with no current limitation

sin limitación de potencia, non-power limited

sincrónica y delta, synchronous and delta

sincronización de fases, phase synchronization

sistema, system

sistema automático de aspersión contra incendios, automatic fire sprinkler system

sistema conectado a tierra, grounded system

sistema de antena, antenna system

sistema de cableado, wiring system

sistema de enfriamiento, cooling system

sistema de generador de reserva, standby generator system

sistema de instrumentación, instrumentation system

sistema de producción de energía eléctrica, interconnected electrical power production

sistema de refrigeración, refrigeration system

sistema de reserva, standby system

sistema de riel de iluminación, light rail system

sistema de señal de miliamperios, milliamp signal system

sistema de supresión de incendios, fire suppression system

sistema de tecnología de la información, information technology system

sistema de toma de agua húmedo, wet-standpipe system

sistema de voltaje/tensión flotante, floating voltage system

sistema Delta de cuatro cables, four-wire Delta system

sistema derivado por separado, separately derived system

sistema ecualizador pasivo, passive equalizer system

sistema eléctrico integrado, integrated electrical system

sistema fotovoltaico solar, solar photo-voltaic system

sistema sin puesta a tierra, ungrounded system

sistema subterráneo metálico local, local metallic underground system

sistema trifilar monofásico, single-phase three-wire system

sistemas de cables Romex, Romex cable system

sobre el nivel del suelo, above ground

sobrecalentamiento, overheating

sobrecalentamiento bajo carga, overheating under load

sobrecalentar, overheat

sobrecarga, overload

sobrecarga térmica, thermal overload

sobrecorriente, overcurrent

sobresallente, recessed

sobretensión, overvoltage

sobretensión transitoria, surge

sodio de alta presión (Lucalox), high pressure sodium (Lucalox)

soga tiracables, cable pulling rope

soldadura, solder

soldadura de arco, arc welding

soldadura de gas, gas welding

soldadura exotérmica, exothermic welding

solenoide, solenoid

sólido, solid

sólo para uso portátil, for portable use only

sometido a carga, under load

soplete de corte, cutting torch

soplete de propano, propane torch

soporte, support

soporte de accesorios, support of fixtures

soporte de alambre, wire support

soporte Minerallac, Minerallac support

subida de voltaje, voltage surge

subíndice, subscript

subpanel, sub-panel

subterráneo, underground

suelos expansivos, expansive soils

suministro, supply

suministro de energía, power supply

suministro de energía no interrumpible, uninterruptible power supply

superficie de adaptación de máquina-tierra, machine-ground matching surface

superficie en contacto con el suelo exterior, surface in contact with exterior soil

super-pegamento, super glue

surfactante, surfactant

sustituto, replacement

T

T horizontal, horizontal Tee

T vertical, vertical Tee

tabique de mampostería huecos, interior hollow masonry partition

tabiques, partition

tabiques cableados reubicables, relocatable wired partitions

tablarroca, sheet rock

tableros de distribución, switchboards

tamaño, size

tamaño comercial, trade size

tanque, tank

tanque de agua presurizado, pressurized water tank

tapa del tomacorriente, attachment plug cap

tapas a prueba de intemperie, weatherproof covers

tarjeta de captura, capturing board

tarjeta de circuitos, circuit board

tarjeta de circuitos impresos, printed circuit board

tarjeta de extracción, removing card

techo con una barra en T, T-bar ceiling

techo estructural, structural ceiling

teclado numérico, numeric keypad

técnico en electrónica, electronics technician

temperatura ambiente, ambient temperature

temperatura diferencial, differential temperature

tendido de cables, wireway

tendidos de los conductos, conduit runs

tensión, strain

tensión superficial, surface tension

térmico con nylon, thermal-with nylon

terminal, terminal

terminal de orejeta doble, double lug

terminal del accesorio, fixture terminal

terminales de la batería, battery leads

terminales de orejetas, terminal lugs

termistor, thermistor

termocontracción, heat-shrink

termoeléctrico, thermoelectric

termoestable a prueba de agua, thermoset-waterproof

termopar, thermocouple

termoplástico resistente a la humedad, moisture-resistant thermoplastic

tiempo de respaldo/reserva, back-up time

tiempo de retardo, on time delay

tiempo de transferencia, transfer time

tierra-arranque, ground-start

tijera eléctrica, electric hedge trimmer

tipo apto para conexión a tierra, grounding type

tipo de fijación, locking type

tipo no apto para conexión a tierra, ungrounded type

tira de bornes, terminal strip

tiracables, cable puller

tiristor, thyristor

tiristor de potencia, power thyristor

toma múltiple, multi-tap

tomacorriente, outlet

tomacorriente de tipo de tierra aislado, isolated ground type receptacles

tomacorrientes estilo "California", California style receptacles

tornillo, screw

tornillo autorroscante, sheet metal screw

tornillo de fijación, set screw

tornillo para madera, wood screw

tornillo plateado, silver screw

tornillo Tek, Tek screw

torque, torque

torsión, twisting

transductor, transducer

transferir, transfer

transformador, transformer

transformador, transformer

transformador con circulación de aceite, oil-filled transformer

transformador de potencial, potential transformer

transformador de tipo seco, dry type transformer

transistor, transistor

transistor de efecto de campo, field-effect transistor

transistor de unión bipolar, bipolar junction transistor

transitorio, transient

transmisión de energía, power transmission

transporte, transportation

trayecto del conducto, conduit routing

trazado conductor, conductive path

trazado de secciones, sectional drawing

trazado separado, separate path

trazador TIC, TIC tracer

trenzado, stranded

tres-estados, tri-state

tridimensional, three-dimensional

triplay, plywood

troquelador de caños, pipe die

troquelador de conductos, conduit die

tubería de gas subterránea, metal underground gas piping

tubería eléctrica de metal, electrical metal tubing

tubería eléctrica metálica (EMT), electrical metallic tubing (EMT)

tuberías eléctricas no metálicas (tubo Smurf), electrical nonmetallic (Smurf Tube) tubing

tuerca, nut

tuerca de resorte, spring nut

tuercas para alambre, wire nuts

U

unidad autónoma, self-contained unit

unidad de disparo, trip unit

unidad de procesamiento de datos, data processing unit

unidad multifamiliar, multifamily unit

uniones roscadas, threaded joints

V

vaina externa de cobre o acero, outer sheath of copper or steel

válvula de prueba de inspección, inspector's test valve

vapores corrosivos, corrosive vapors

variable, variable

varilla, rod

varilla de puesta a tierra, ground rod

varilla de tierra conducida, driven ground rod

varillas de materiales no ferrosos, rods of nonferrous materials

varistor, varistor

vataje, wattage

vataje total de las lámparas, total wattage of the lamps

vatímetro, wattmeter

vatio, watt

vatios por pie lineal, watts per lineal foot

vatios-hora, watt-hours

velocidad de base, base speed

verificación por instrumentos, instrument scan

verificar, check

vibración, vibration

vidrio resistente a la temperatura, temperature resistant glass

viga, beam

visor microscópico, microslide viewer

vista transversal, cross-sectional view

voltaje a tierra, voltage to ground

voltaje del circuito, circuit voltage

voltaje del circuito abierto, open circuit voltage

voltaje en condiciones de carga, voltage under loaded conditions

voltaje máximo, peak voltage

voltaje primario, source voltage

voltaje/tensión, voltage

voltaje/tensión de carga, load voltage

voltaje/tensión de CD, DC voltage

voltaje/tensión de línea, line voltage

voltaje/tensión nominal, nominal voltage

voltamperios, volt amperes

voltamperios, volt-amps

voltímetro, voltmeter

voltio, volt

Y

yugo, yoke

Z

zapata, footing

zinc, zinc

ABOUT THE AUTHOR

For more than 25 years, José Luis Leyva has worked as a translator and interpreter in various technical areas. His vast experience in bilingualism has allowed him to interpret for CEO's, manufacturing directors, human resources managers, plant managers, attorneys, ambassadors and even Presidents. He is also the author of other books, including technical dictionaries of the *The 1333 Most Frequently Used Terms* series.

ACERCA DEL AUTOR

Durante más de 25 años, José Luis Leyva se ha desempeñado como intérprete y traductor en diversas áreas técnicas. Su amplia experiencia lingüística lo ha llevado a interpretar para directores ejecutivos, directores de manufactura, gerentes de recursos humanos, gerentes de planta, abogados, embajadores y hasta presidentes. Es también autor de varias obras, entre las que se incluyen los diccionarios técnicos de la serie *The 1333 Most Frequently Used Terms.*

www.ingramcontent com/pod-product-compliance
Lightning Source LLC
Chambersburg PA
CBHW051508170526
45166CB00001B/444

* 9 7 8 1 4 9 0 9 4 9 2 4 6 *